Héritier KYABENE MAKONGA

Comparative study of group and individual assessment

Héritier KYABENE MAKONGA

Comparative study of group and individual assessment

The case of 6th grade students at the Imani Panzi Institute

ScienciaScripts

Cover image: www.ingimage.com

This book is a translation from the original published under ISBN 978-620-6-71964-9.

Publisher:
Sciencia Scripts
is a trademark of
Dodo Books Indian Ocean Ltd. and OmniScriptum S.R.L publishing group

120 High Road, East Finchley, London, N2 9ED, United Kingdom
Str. Armeneasca 28/1, office 1, Chisinau MD-2012, Republic of Moldova, Europe
Managing Directors: Ieva Konstantinova, Victoria Ursu
info@omniscriptum.com

Printed at: see last page
ISBN: 978-620-8-53326-7

EPIGRAPH

Behold, how good and how pleasant it is for brothers to dwell together!

Psalms 133:1.

The Holy Bible, Darby.

Coming together is a start; staying together is progress; working together is success.

Henry Ford

To the Creator of the Universe, Almighty God, Alpha and Omega, he who strengthens us and who by his Grace enables us to do all things; given the greatness of your benefits, we exalt you Lord

To you,

We dedicate this work.

IN MEMORIAM

Over the past few years, we have lost loved ones to whom we dedicate this page:

- *Dearest parents remembered by God, Papa Valentin KITAKUBILI MAKONGA and Maman Georgette MITUNI KATAWANDJA,*
- *Little brother Fortune MAKONGA and little sister Emérence MAKONGA,*
- *Uncles Bernard KYANDE MOLIGI and Issa KAMPENGE KATAWANDJA,*
- *Grandparents Alexis MOLIGI KITAKUBILI, Clément KATAWANDJA KILIMO, Shelina MUMBYA FERUZI and Rose NYANDWI.*

Despite having left us so soon, we hope to see you again one day by the Grace of God, and now we say:

Peace be to your souls!

ACKNOWLEDGEMENTS

Man being a social animal, it was not easy to do this work alone. So it is to show our gratitude to those who helped us that we dedicate this page. First of all, to the academic authorities of USK/Bukavu in general, and those of FPSE in particular, for all their efforts. To CT. David Malala Ntambue and Gérard Mubangu Wa Kapala, respectively dictator and co-director of the present work, who did their utmost to provide us with observations and guidelines, without which this work would have no meaning or scientific value.

To Mr. Méschac Vunanga, the provincial coordinator of ECP/SK and his team, for giving us the opportunity to spend our internship at the aforementioned structure, where we were able to obtain information about the PAP from the source.

To the management of the Institut Imani Panzi, to the teachers of this school and particularly to Messrs Isaac Bwami and Marcelin (French teachers) and also to the students of 6ème MP and CA 2016 edition.

Last but not least, we are indebted to our family and friends who, in one way or another, have supported us in the realization of this work, and brought us light when we needed it most.

As your list is so long, we would like to express our sincere thanks to all those who have contributed to the success of our academic career in general, and this work in particular. God alone will reward you for all your efforts.

May Almighty God bless you!

TABLE OF CONTENTS

GENERAL INTRODUCTION

Hanna Dumont, David Istance and Francisco Benavides (2010, p. 2) point out that today we need to *rethink what is taught, how it is taught, and how learning is assessed*.

To answer the question of how it's taught, Unicef (2009, p.23) argues that it'*s child-centered, interactive methods that make learning fun and exciting for students, and improve retention, participation and achievement*. These methods facilitate the creation of open learning environments characterized by intra-group cooperation and positive competition between learners. These new methods transform the teacher from a *source of all knowledge* and a feared authority figure into a *facilitator learning* and someone who listens to students. In this way, teachers encourage learners to take responsibility for their own learning, so that the motivation to learn comes from the learners themselves rather than being imposed from the outside. In other words, far from being simple observers, students have become true actors in their own learning, especially as they actively participate in the teaching-learning process.

Addressing the question of the *effectiveness* of education and/or schooling, Lumeka, quoted by Gratien Mokonzi (2006, p.8), believes that an effective school is one that provides pupils with basic knowledge, teaches them to teach themselves, nurtures their creativity and supports their whole being. Reinforcing this point of view, Delors *et al.* quoted by Mokonzi (Ibid.) specify that a school is effective when it *"provides training that enables the individual to discover, awaken and strengthen his or her creative potential...a school that enables each individual to better understand his or her environment, in its various aspects, encourages the awakening of intellectual curiosity, stimulates the critical sense and enables the deciphering of reality by acquiring autonomy of judgment"*.

To this end, Delors *et al* (1998. pp.83-84) spoke of four fundamental apprenticeships that constitute the *pillars of knowledge* which, once applicable in a pupil's life, will make his or her school effective: :

- ✓ *Learning to know* (acquiring the tools of understanding) ;
- ✓ *Learning to do* (to be able to act on one's environment) ;
- ✓ *Learning to live together* (to participate and cooperate with others in all human activities and, ultimately, to *live in harmony with* others).

✓ *Learning be* (which is an essential path that contributes to the previous three). They add that these pillars are inextricably linked, as there are multiple points of contact, overlap and exchange between them.

To paraphrase Micheline Bercier-Larivière and Renée Forgette-Giroux (1999, p.169), traditional pedagogy, characterized by the transmission of knowledge and the selectivity of the best students, is gradually disappearing. As a result, add Micheline Bercier-Larivière *et al* (Ibid., p. 177), evaluation has evolved and adapted to the new philosophies of the education system, to the new teaching and learning theories it adopts, and to the missions it sets itself. Later, in an effort to make assessment transparent, these authors emphasize participation, discussion and appropriation by students of the various aspects of the assessment of their learning.

For Gratien Mokonzi (2009, p. 121), assessment is one of the five competencies expected of a professional teacher in the classroom. We would add with this author (2015a, p. 3) that assessment is a valuable tool in the teaching-learning process. This makes all the more sense given that, as Unesco (2014, pp. 279 and 287-288), Joutard and Thélot and Marc Romainville (2002, p. 43) point out, evaluation is the "*mirror effect*", i.e., it enables those involved in the teaching-learning process to clearly realize what they have just done. In other words, by acting as a mirror, evaluation enables teachers and learners to have a precise or clear view of what they have done, so they can modify, if necessary, their practices to make them more effective.

To this end, Abernot, as quoted by Nasser Aboubaker (2009), stresses that, at the end of each learning task, assessment should be used to inform the parties involved in the teaching-learning process about the level or degree of mastery achieved by the student, and to help both parties discover where and how the student is still experiencing difficulties. This helps us to understand that assessment helps to optimize the teaching-learning process, especially as it makes visible the strengths and weaknesses of the aforementioned process, so that other strategies can be devised by both players to meet the challenges.

As assessment is at the service of the teaching-learning process, it is clear that it must ipso facto be underpinned by the pedagogical method that drives the process in question. In line with what has just been said, and taking into account the opinions supported by many players involved in the field of education and children, including Unesco (2005 and 2006), Unicef (2009) MINEPSP (2010 and 2012) researchers such as Gratien Mokonzi and Christian Grêt (2006 and 2009), etc., for whom the application of pedagogy advocating active participation, cooperation or collaboration... is an imperative today to enable schools and children to become more effective. for whom the application of a pedagogy based on active participation, cooperation or collaboration... is nowadays an imperative to enable schools and children to become more efficient.

From this perspective, Aline Germain-Rutherford of the University of Ottawa writes as follows: *If you want to change something in your students' learning process, then change your assessment methods*

The Ministère de l'éducation, du loisir et du sport de Québec (2006, p. 46) proposes the following evaluation practices to promote skills development:

- Focus on regulation;
- Encourage student collaboration and
- Use appropriate assessment tools

Gérard Mubangu (2015, p. 159) points out that the selective, individual nature of assessment is gradually disappearing wherever the PAP is applied, in favor of group assessment.

The system individual assessment that characterizes Congolese schools in general, and those in the city of Bukavu in particular, calls for a comparative study between individual and group assessment, to identify which contributes more to learners' results in schools. Observation of these two types of assessment leads us to ask the following question: *does group assessment have a significant influence on the results of secondary school pupils in Bukavu?*

In light of the above question, the following answer is predicted: *the practice of group assessment has a significant influence on learners' academic results.* In other words, the *application of the socio-constructivist model in the assessment of students enables them to do better than when they work alone or individually.*

The choice of this topic is based on the fact that *assessment of prior learning* is a crucial element in the teaching-learning process. Having nowadays two forms in its application in the classroom, it's important to compare the two to find out which is more effective than the other.

Let's just say that the interest of this work is twofold: firstly, it enables the student researcher to combine the theories learned not only in relation to research methodology, but also (and above all) those of other disciplines that are relevant to the subject being dealt with. In addition to this dimension, this work is also a contribution to the

improvement of the assessment system in Congolese schools in general, and in Bukavu in particular.
In short, over and above the practical application of the various notions learned in the auditorium, this work gives spectacles to all those involved in the educational sector, enabling them to realize the performance of group work in the process of assessing student learning.

Generally speaking, to undertake this study is to understand how effective the practice of PAP is in the process of assessing students' learning achievement. Or, quite simply, to understand the influence that the application of PAP in assessment can have on learner performance.

Specifically, once the hypothesis formulated has been corroborated, the aim is to understand at what level this pedagogy influences the process of assessing learning achievement. This will enable us to verify, at our level, the necessity of the PAP that other education specialists attribute to this new pedagogical approach. In short, the aim is to make clear the importance of PAP today's evaluation practices in Bukavu schools first and foremost, but also throughout the DRC if possible.

To achieve this, the documentary technique and the experimental method (quasi-experimentation) helped the researcher to collect the various data that are the subject of this work. As for the selection of the sample, stratified random sampling was used.

Finally, to process the data and quantify the results, the indices of central tendency (in the case of the mean), dispersion (variance, standard deviation and CV) and yield are calculated before the comparisons of the means are made. These comparisons are made using Dunett's t-test.

Having dealt with group work as a whole is difficult, if not impossible, given the time available for this research. In addition, financial constraints,... and other reasons not mentioned above, meant that this theme was dealt with in only a few specific aspects (space, time and theme). Thus, from the spatio-temporal point of view, this work was carried out in the Protestant educational sector of the province of South Kivu for the 2015-2016 school year precisely at the Institut Imani Panzi in the two classes of 6th secondary. And finally as for the thematic delimitation, the MPAP or PAP and traditional pedagogy (competitive, selective,...) are confronted in their application of the process of evaluation of students' achievements.

Apart from the introduction and conclusion, the work covers four chapters:

- Conceptual and theoretical points of reference, in which *the key concepts of* the work (evaluation, constructivism, socio-constructivism, quality education and the teaching-learning process) will be clarified; *pedagogical models*; an *overview of the PAP, forms of assessment of learning achievement,* and finally *some previous works* that have influenced the present work will be presented in a few lines;
- Methodological frameworks in which full details are given of the application of various methods and techniques to enable the study to achieve its objectives;
- The presentation and processing of data and interpretation of results, which includes a point on the presentation of the data collected and another on the processing or analysis of this data and the interpretation of the results of the analysis, and finally
- Discussion of the results, in which the results obtained are compared and/or supported by the ideas of the great educators, pedagogues, etc.

Chapter 1: CONCEPTUAL AND THEORETICAL BACKGROUND

In this chapter, five points are developed: definition of key concepts, overview of pedagogical models, PAP, assessment of learning achievement and, finally, some previous studies are presented.

1.1. DEFINITION OF KEY CONCEPTS

Let's start by pointing out that a word can be defined in different ways, depending on the field of research. It is worth noting that the few key words in our theme will be defined within the framework of the educational sciences.

1.1.1. Evaluation

For Jean Mari de Ketele, quoted by Franc Morandi and René la Borderie (2006, p.121), *evaluation means making a value judgment on the adequacy of a set of information and a set of criteria in relation to a decision-making objective.* These authors point out that evaluation can cover a wide range of fields, including *student performance, the effectiveness of a training program or teaching method, assessment of knowledge at national (*TENAFEP and EXETAT*) or international (*for comparative education, PASEC, for example) *level, and also the operation of the education system.* Renchérie par Françoise Campanale (2001, p.2), evaluation is not only concerned with a product (a student's achievement), it is also focused on the process (approach used).

1.1.2. Constructivism

For Henriette Bloche *et al* (1999, p.209), this is a theoretical position on ontogeny that views development, whether biological, psychological or social, as the construction of relatively stable given organizations that follow one another over time. Constructivist theories are theories of structural development.

1.1.3. Socio-constructivism

Following on from the constructivist movement, social constructivism, developed by Lev Semenovich Vygotsky, incorporates, as its name suggests, the *social dimension*. The socio-constructivist perspective emphasizes the role of multiple social interactions in the construction of knowledge, and proposes that learning be seen as active participation in real-life activities, interacting with others. Socio-constructivism is therefore based on the following principles:

- ✓ The student's head is never empty of knowledge;

- ✓ Learning does not happen by piling up knowledge, nor in a linear fashion;
- ✓ Social interaction between students can help learning;
- ✓ Students give meaning to knowledge if it appears to be an indispensable tool for solving a problem.

1.1.4. Quality education

Clearly, the concept of *quality education* is difficult and sometimes impossible to define, especially as it is complex and varies according to the conditions and expectations of the context in which it operates. For example, Unesco's Framework for Diagnosis-Analysis of Quality in General Education, quoted by the European Youth Forum (2013, p. 6), has this to say on the subject: *the definition of quality education is intrinsically linked to an understanding of the purpose of education in a given society, and in view of the existing development needs and aspirations of the individual and the community.* This source (Ibid., pp. 6-7), adds that quality education embodies the following principles: *accessibility, equity and inclusion, community impact, learner (student) participation, parity and reciprocity in the educator-learner relationship, cooperation and complementarity, and ultimately support.*

Thus, Anastassis Kozanitis (2005a, p. 11), writes that quality education is one that deals with *what to do to learners* and *how to learn* (showing how to learn).

Beyond the two questions proposed by Kozanitis, we project other essential aspects that come under the 5W+1H rule, which is often used by journalists, as André Giordan and Jérôme Saltet (2011, p. 23) point out, according to which, for 5W, we have to ask questions (Who?, What?, When?, Where? and Why?) and 1H comes back to the question How?

In short, for us, defining *quality education* means responding favorably or positively to the six questions listed above.

1.1.5. Teaching-learning process

By teaching-learning process, we mean a dialectical sequence of operations involving a combination of teaching efforts on the one hand, and those of the learner on the other, in order to achieve the pedagogical objectives pursued by a given didactic sequence.

1.2. THEORIES ON PEDAGOGICAL MODELS

By pedagogical model, Franc Morandi *et al* (op.cit, p.126) mean a coherence between educational objectives, a conception of knowledge and relations between teacher and pupils. Based on these models, school achievements (teaching and learning) propose *methods* that bring together actors and define the material, cognitive and social characteristics of a pedagogical practice.

They add that pedagogical models are not "instructions for use", but solutions chosen to achieve joint work between student and teacher. These are the possible logics for *learning* and *making learning happen.*

For Anastassis Kozanitis (op.cit, p.1), the theoretical foundations of the educational sciences are to be found in psychology, sociology, philosophy and cognitive science, among others. This diversity of theoretical fields underpins the different approaches to teaching and learning, and can sometimes be confusing insofar as certain authors can be found within more than one theoretical stream. Gabriel Labédie and Guy Amossé (2013) and the IREM de Toulouse (n.d, p.1), in an effort to answer the question of *how to promote learning for the greatest number of students, without hindering future learning, in the time available,* help us to list three major currents or pedagogical models that apply to the teaching-learning process. These models are: the *transmissive* model, the *behaviorist* model and, finally, the *constructivist* models.

1.2.1. The transmissive model

As IREM de Toulouse points out (Ibid.), this model, inherited from traditional pedagogies, reduces *learning to the recording in memory of the knowledge presented by the teacher*, as if this knowledge were printed directly in the student's brain like photographic film. The act of teaching is central to this model, since it is the teacher who expresses, demonstrates, constructs and structures knowledge. There's nothing to learn when the teacher doesn't speak or show.

The student simply listens attentively and receives the knowledge in his or her supposedly empty head. This knowledge is shaped from the outside, and must be adapted to the teacher's lecturing and/or questioning activities, in a situation of collective, vertical communication.

The teacher is not allowed to make any mistakes in his or her presentation, and even when he or she asks a question or questions, the student must give the right answer to show that he or she has fully understood. In short, in this model, the student is in a situation where two verbs are conjugated in the present tense and in the first person singular, i.e. "*I learn/apply*". In other words, once he's followed what the teacher says, all he has to do is reproduce it as it is, to show that he's understood.

Schematically, the student is understood as follows:

Figure 1. The transmissive model

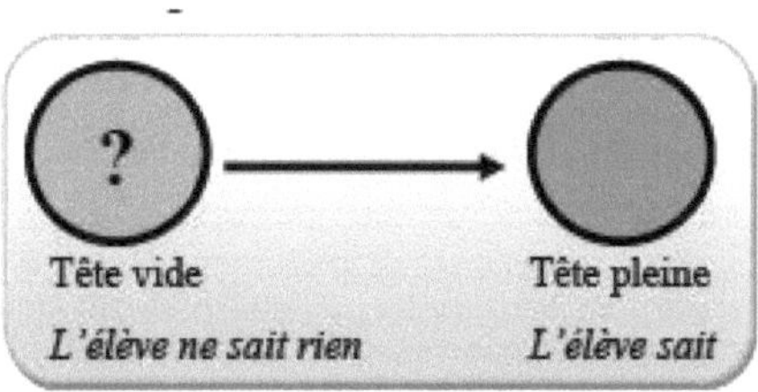

Source: Gabriel Labédie *et al* (op.cit)

1.2.2. The behaviorist model

For Henriette Bloche *et al* (op.cit, p. 81), this model conveys the idea that acquisitions are the product of external events, with the individual passively undergoing the process. This trend can be traced back to the psychological work of John B. Watson in the USA, and was later extended to the analysis of human learning and the field of education.

In this model, learning is defined as the *ability to respond appropriately to given stimuli*. And in this sense, learning is seen as a mechanical process in which the learner's behavior is determined by the reinforcements encountered: *good responses* are rewarded and *bad ones* are punished and abandoned. In short, it's *learning by conditioning*, which tends to suppress learners' mental activity.

For Gérard Barnier (n.d., pp. 5-6), behaviorists consider mental structures to be like a black box to which we have no access, and that it is therefore more realistic and effective to focus on *"inputs"* and *outputs"* than on the processes themselves. This author adds that the behavior these behaviorists are talking about is not an attitude or a way of being on the part of the student, but rather the usual meaning of the word when we say that he must improve his behavior. This implies that the student must demonstrate positive behavior after a didactic sequence.

In addition to the work of J.B. Watson, the work of Skinner, Thorndike and others has reinforced this pedagogical model. Let's say that the student's perception of this model is schematized in the figure below:

Figure 2. The behaviorist model

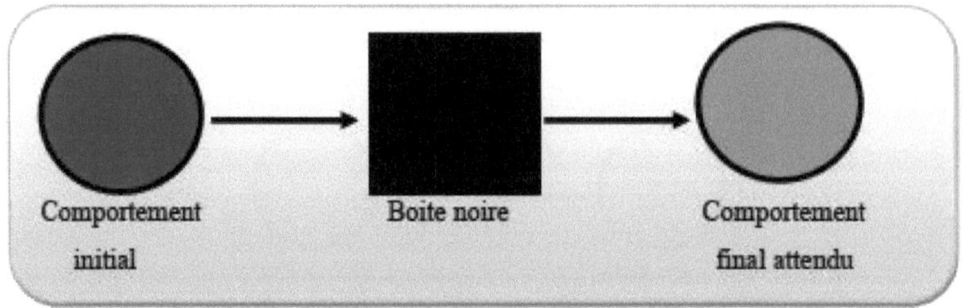

Source: Gabriel Labédie *et al* (op.cit)

1.2.2.1.*Jean Piaget's individual constructivism*

All we need to remember in relation to this current is that the learner is the actor or constructor of his or her knowledge, with the support of the environment and his or her biological growth. For Piaget, as the Toulouse IREM explains (Ibid., p.4), *assimilation* and *accommodation* constitute two poles of adaptation that are inseparable if knowledge is to be constructed. In addition to these aspects, Isabelle Girault (2007, p.18) adds that development is characterized by the passage from one structure to another through the process of *equilibration*. Or, to put it another way, equilibrium is achieved through assimilation and accommodation.

Schematically, this involves :

Figure 3. Jean Piaget's adaptation

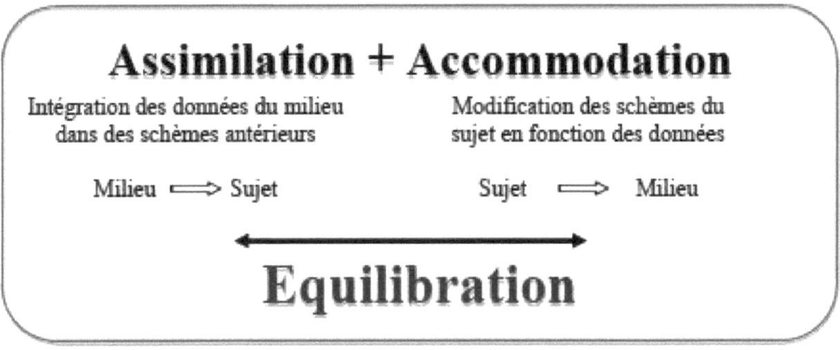

Source: Isabelle Girault (op.cit, p. 19)

1.2.2.2. *L. S. Vygostki's social constructivism or socio-constructivism.*

In addition to the individual or personal aspect advocated by Jean Piaget, Vygotsky, as described by Henriette Bloche *et al* (op.cit, p. 936), adds that the child needs to be in a social setting to construct knowledge. Thanks to his work, other researchers later reinforced this social approach to constructivism. These authors include : Perret-Clermont, Doise and Mugny quoted by Joshua and Dupin, and taken up by Anastassis Kozanitis (op.cit, p. 12); Bruner quoted by Gabriel Labédie *et al* (op.cit) and many other researchers.

What's interesting is that, as Céline Buchs, Katia Lehraus & Fabrizio Butera (2006, p. 177) point out, all these researchers approach the social constructivist approach where interaction is conceivable, and argue that *exchanges* and *discussions between students* in interactive learning situations foster their dignified learning. Régis Le Coultre (n.d, p. 1), for example, seeks to provide an answer to the question posed *by the social constructivist model: "How is knowledge constructed?* He points to the interactionist approach, with Doise and Mugny as emblematic figures of this approach, which follows on from the work of Vygotsky. In view of this, he concludes that what's important is that in all these theories, there are several members collaborating simultaneously.

Régis Le Coultre (Ibid., p. 2) proposes the following figure to illustrate this point:

Figure 4. The construction of knowledge in socio-constructivism

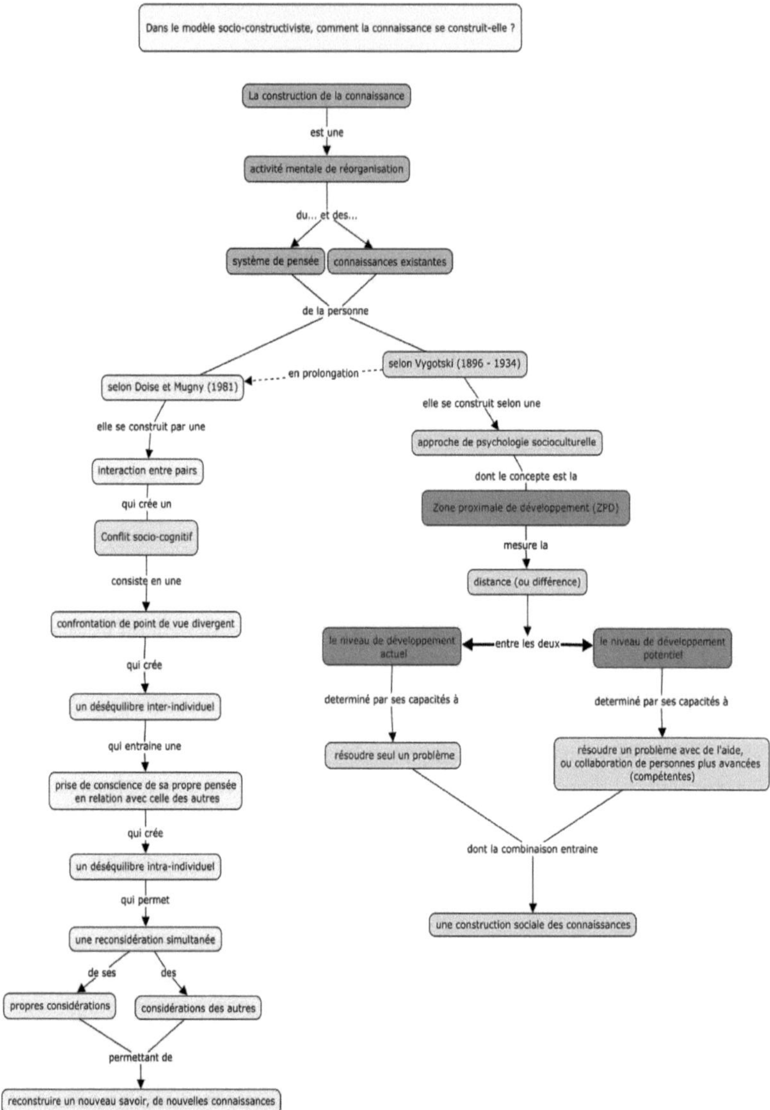

Aware that *the interactive situation* in the classroom where students confront are called upon to complement each other to achieve the objective, Céline Buchs *et al* (op.cit, pp. 189-190) argue that there are confrontations that are not beneficial to the teaching-learning process. However, to organize effective group work, say Céline Buchs, Laurence Filisetu, Fabrizio Butera and Alain Quiamzade 2004, p. 180), the teacher as guide must :

- Propose a common task that can be carried out as a group: if this can be done individually, there's no reason for students to interact;
- Getting a small number of students to work together;
- Offer activities that build team spirit and trust: it's essential to create a positive climate before starting the actual work, with activities that foster knowledge of others, respect and support among participants;
- Create positive interdependence between students: the common goal must be structured in such a way that each member of the group perceives that he can only achieve his goal if all the members of his group also achieve theirs;
- Maintaining the personal responsibility of each group member: ensuring that each participant does his or her best to accomplish his or her share of the work;
- Teach the social and interpersonal skills needed for group work: it's not enough to want to cooperate to know how to do it. Determine a useful social skill, then work on how to put it into practice (verbal or non-verbal);
- Get students to reflect on their academic and social skills: this reflection focuses on what has been useful to move the group forward and what can be improved and, at the end of the day, what can be done better.
- Reinforce cooperative behaviors and constructive interactions: the teacher observes and gives positive feedback on valued behaviors.

In this sense, Smith, Johnson David and Johnson Roger quote Fadi El-Hage (2013, p.1), for cooperative learning to take place, each member must contribute to the learning of others; take on their share of the work and practice the skills required for cooperation to be effective.

Thus, according to Anastassis Kozanitis (op.cit, p.12), socio-constructivism requires three didactic elements to enable the individual to progress:

- The *constructivist* dimension: alluding to the learner himself,
- The *social* dimension: involving classroom partners (colleagues and teacher),
- The *interactive* dimension or *environment*: situations and the learning object (teaching content) organized within these situations.

This brings us back to the situation where the learner must be placed in a situation in which he must confront what is inside him and what others know about, in order to find a final solution. Schematically, we have the following:

Figure 5. Constructivist models

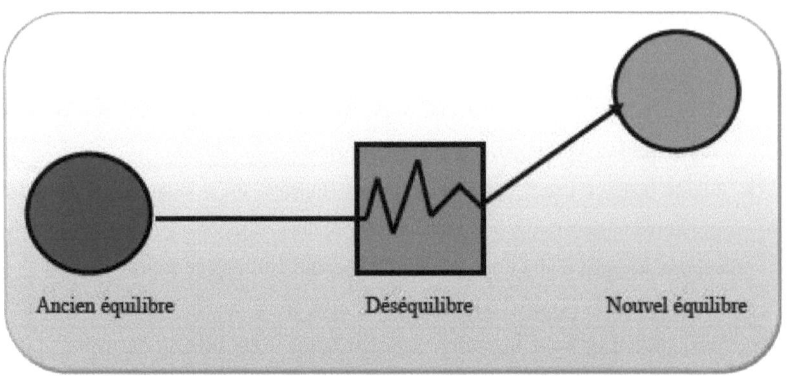

Source: Gabriel Labédie *et al* (op.cit)

The table below outlines the contributions of these two eminent constructivist psychologists.

Table 1. Contributions of constructivist psychologists.

Jean Piaget	Lev S. Vygotski
Acquisition is construction	**Acquisition is appropriation** It's the social significance of objects that matters. The subject alone in the world could learn nothing at all.
The role of language in development knowledge is	**The role of language** in the development of knowledge is crucial

Development precedes learning (Mentalist conception)	**Learning drives development.** Vygotski distinguishes two situations: ✍ The one where the learner can learn and perform certain activities alone, ✍ The one where the learner can carry out an activity with the support of another. This determines the learner's *potential for development.* Between these two situations lies the ZPD, in which the individual can progress thanks to the support of others.
Pedagogy of discovery: the child makes The results of these experiments are subtle and interesting.	**Mediation pedagogy**: the mediator intervenes between the child and his environment. In a given culture, children cannot rediscover everything themselves.

Source: Gabriel Labédie *et al* (Ibid.)

The lesson drawn from these two theories is what Christian Grêt (n.d, op.cit) points out when he says that the Piagetian individualist and Vygotskian social approach to the appropriation of knowledge at the end of the 20th century favored the emergence of *self-constructivism*, the latest mode of learning, after imprinting or imitation (transmissive) and conditioning (behaviorist).

To simplify matters, Christian Grêt (2009, p. 148-149) explains the three pedagogical models in the following table:

Table 2. Pedagogical models

Models	Student roles	The teacher's role	Error status	Knowledge position	Model limits

Transmissive	Listen, pay attention and don't search	Communicating and demonstrating knowledge	Mistakes must be avoided	Speaker to receiver	You learn by paying attention. No motivation
Behaviorism	Solve tasks prepared by the teacher	Communicating or demonstrating knowledge	Mistakes must be avoided. They leave their mark.	Knowledge is discovered by the pupil according to a path predetermined by the teacher.	No initiative on the part of the student. Motivation problem
Constructivists	Construct knowledge by being put in a problem situation	Propose problem situations. Lead the results comparison phase.	Mistakes are reconsidered here, as they can be used to raise awareness of new concepts.	Knowledge is built by the student	Michel Chastelin says: this model is time-consuming. Can everything be handled this way?

As for what has just been said, and paraphrasing Gérard Mubangu (2014, p. 160), it will be important to merge the transmissive and behaviorist models into the camp of the *traditional pedagogical* approach, especially as they all envisage the pupil as an object on whom the teacher does whatever he or she pleases. In short, these models place particular emphasis on the teacher, to the detriment of the student, who should be considered the main actor. And on the other hand, Piagetian constructivism and Vygotsky's approach have formed a second camp, the *PAP*, which sees the child as the main actor in the construction of his or her knowledge through his or her peers and the teacher as the guide. This is clearly illustrated in the following table:

Table 3. Summary of pedagogical approaches

PAP approach	Traditional teaching approach
➕ Constructivism Students build their own knowledge, know-how, interpersonal skills and future skills... ➕ Socio-constructivism Students build knowledge, know-how, interpersonal skills, future skills, etc. in groups: mutual exchange of experience and knowledge.	➕ Transmissive Magisterial character: imitation, memorization of knowledge, tabula rasa, companionship, dogmatic, ex-cathedra method, the master's lecture. ➕ Behaviorist Conditioning: behavior modification
The learning paradigm	The teaching paradigm

Source: Gérard Mubangu (op.cit, p. 160)

1.3. PAP OVERVIEW

Before going into more detail, we should first point out with Daniel Kambale (2015, p. 1) that this pedagogy derives from Jean Piaget's *auto-constructivism* (individualist model) and Lev Semenovitch Vygotsky's *socio-constructivism* (social current), for which, without contact with others, learning remains embryonic.

What's more, it's worth mentioning its entry into the DRC, which, according to Gratien Mokonzi (2006, p.2) only dates back to 2005, when the very first training session was held for provincial and community coordinators in Kinshasa. And in the province of South Kivu, as Mr. Méschac Vunanga points out in a paper on the evaluation of the *"promotion of school management and PAP training* project, it was only in 2011, at the request of the CP-ECP/SK, that the very first session was held for teachers and various educational partners under the latter's management. In this communication, we note that since the introduction of this approach in the province, 630 educational players have so far benefited from training in this approach, the majority of whom work in the Protestant sector, including teachers, headteachers and educational advisors, as well as some educational inspectors.

1.3.1. What about PAP

For Christian Grêt (2006b, p.10), the PAP is a trend that advocates the child's active participation in his or her own education. The essential innovation brought about by this pedagogy is that it: "seeks to make the learner *active,* it starts from his or her centers of interest, strives to awaken *cooperation* rather than competition, and favors *discovery* over exposition".

He adds that this pedagogical approach places learners in problem situations, enabling them to construct their own knowledge. The learner is involved in situations that enable him or her to use skills and develop them over the course of the course. As a result, the teacher's role changes radically from that of the transmissive style; he or she encourages the students' research and facilitates their confrontations when working in groups. In short, with this approach, the teacher no longer gives lessons, but rather organizes the learning patterns that enable students to work on and develop their knowledge.

The two qualifiers (active and participative) underline the fundamental dimensions of modern pedagogy. It is *active* because it sees the student as the driving force behind his or her own learning, through the activities he or she carries out to acquire knowledge, with the support of the teacher, who becomes only a *guide.* It is also *participatory*, as it encourages children to take part in the construction of their own knowledge. It's worth pointing out that encourages *group work*, making the child a cooperative person.

Taking into account Jean Houssaye's reflections (1992, pp. 40-41) according to which: "every pedagogical situation seems to us to be articulated around three poles, including *knowledge-teacher-students*, but, operating on the principle of the excluded third, the pedagogical models that emerge are centered on a privileged relationship between two of these terms. Thus, when a relationship is established between the teacher and knowledge, we're talking about the *teaching* process; when this relationship is established between knowledge and the student, we're talking about the *learning* process; and finally, when it's between the teacher and the student, this process is that of *training.*

For Grêt (n.d), the current trend (PAP) considers that all pedagogical practice is a relationship between three elements at once: knowledge, the teacher and the learner. While expressing regret at those who believe that PAP wrongly places the learner at the center of the teaching-learning process, despite the fact that the learner is only one particle of this triangle, Grêt points out that it is true that the other two components of this triangle (teacher and knowledge) have their own particular and correlative importance, but that this view amounts to demanding that the learner become the real actor in his or her own learning, insofar as the appropriation of his or her own knowledge corresponds more closely to the ideal of socio-constructivist self-learning. To this end, Christian Grêt (2006a, p. 20) adapts the didactic triangle as follows:

Figure 6. Teaching triangle adapted by Christian Grêt

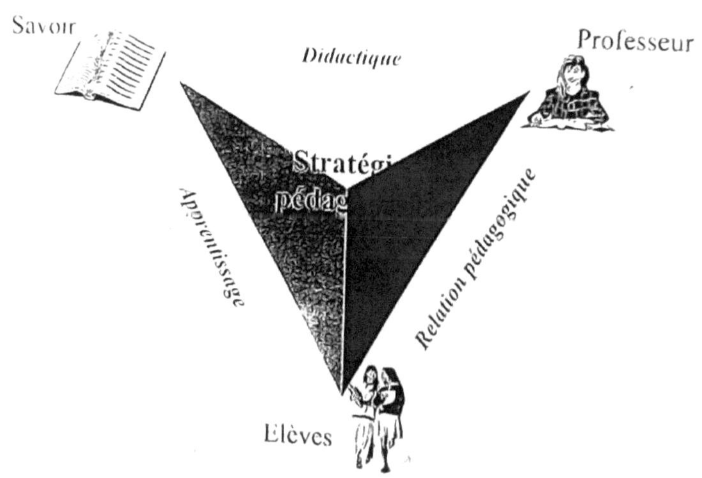

In view of this triangle, the following relationships are established for each teaching-learning process: the *learning* relationship (between knowledge and learners); the *didactic* or *pedagogical strategy* (between knowledge and the teacher) and, finally, the *pedagogical* relationship between teacher and students. For Legendre, quoted by Anastassis Kozanitis (2015, p. 4), this is the set of cognitive, affective and social interactions between learners and teacher, aimed at learning and personal development. In addition, as Gérard Mubangu and Honoré Birindwa M. (2015, p. 129) paraphrase, for Grêt, each of the two players in this process must be involved in specific activities, namely :

- The teacher must :
  - ✓ Manage and lead these groups,

- ✓ Doing a preparation that's quite different from those of other times. Questions like: What are the instructions? What are the learners doing in the groups?
- ✓ Evaluate objectives after each learning sequence, prepare problem situations that are solved by the learners,
- ✓ Facilitate the pooling of learners' results.

❖ Learner:

- ✓ Build your knowledge,
- ✓ He will find the complement of his knowledge in the ideas of his classmates and those of the teacher,
- ✓ They also have to find out for themselves in existing documents (books, manuals, Internet, etc.) what the courses don't offer,
- ✓ The learner is no longer passive, but a researcher...

In this dynamic, triadic system, the student becomes a player in his or her own training. This is more a formative than an informative action. This action is inspired by constructivist theories, which maintain that when trainees take charge of their own training, learning is reinforced.

For Silberman, quoted by the Inspectorate General (2010, p. 3), PAP means giving students the chance to conjugate the following verbs themselves in the first person singular: hear, observe, discuss, do and teach.

- ***What I hear**, I forget;*
- *What I hear and **observe,** I remember a little;*
- *What I hear, observe and **discuss,** I begin to understand;*
- *What I hear, observe, discuss and **do,** gives me knowledge and competence;*
- *What **I teach** another, I master.*

MINEPSP (2012, pp. 26-27), adds that this pedagogy is based on four didactic principles: *activity, participation*, *anticipation* and finally the *child's cooperation.*

- *Activity:* where students' knowledge, attitudes or skills are linked to their needs. It's about teaching students how to learn, make decisions about what they experience and what actions to take.

Drawing on studies by Vayer and Roncin, Dessus (2007) points out that the *activity* is organizing in the group insofar as, if it is well specified, the members of the group (students) will be essentially preoccupied with the defined task, and difficulties will be minor. As the author points out, these studies have shown that :

  - Individuals tend to abdicate their judgment and depend on authors when they find themselves in ambiguous or confused situations;
  - Consensus, or awareness of shared feelings among members of a group, plays a decisive and powerful role in the dynamics of all groups;
  - Conflict and aggression are unavoidable and must be seen as dynamic phenomena, since they require the protagonists to reorganize the information they need to keep the group coherent;
  - If there are common goals, there's a better chance of collaboration within the group;
  - If goals have been set by the group, group members feel more involved than if they have been defined externally;
  - Once the decision has been made to be and do together, the smaller the group, the better it works.

- *Participation*: in this aspect, it's the students who carry out most of the activities. They analyze, study ideas, solve problems and apply what they learn.
- *Anticipation*: this enables students to act for present and future purposes. In other words, students must find that school activities enable them to solve present and future problems.
- *Cooperation*: we talk about this for Dessus (idem), when we're interested in a type of work where students complement each other, and work in small groups. And this enables students to learn together in a complementary and mutual way. Students learn with common goals, receive mutual rewards, use common resources and benefit from complementary roles.

For Hay, quoted by François Le Menaheze (2002, p. 10), there is cooperation in a group when all the members coordinate their actions to achieve a common goal by sharing the various tasks and roles necessary for its realization through coordinated sequences and depending on the collective task to be carried out. Pierre-Yves Cusset (2014, p. 19) adds that cooperative learning is based on working in small, heterogeneous groups. And within the group, students are not left to their own devices: the work is structured to ensure that each student is effectively involved in completing the proposed task. Cooperation can be achieved by encouraging discussion of points of view, or by sharing roles within the group that make students genuinely dependent on each other

The MINEPSP (op.cit, p.25) also points out that this pedagogy is important because it considers the child to be the object, subject and principal agent of his or her education. This is because all educational work is in vain without the child's support and participation. To this end, it helps students to :

- ✓ Develop a spirit of curiosity ;
- ✓ Reinforcing lessons; teamwork;
- ✓ Expand their life's opportunities and purpose;
- ✓ Apply their knowledge to find solutions to personal and community problems;
- ✓ Encourage collaboration and interaction between them, whether girls and boys or students from different religions and tribes.

1.3.2. Methodological procedure for PAP lessons

Despite the complexity of this didactic tool in terms of its applicability in the classroom, Christian Grêt (n.d, op.cit) and Gratien Mokonzi (2006, pp. 4-5) propose four main stages: *task presentation*, *individual and/or group work*, *restitution or pooling* and *synthesis.*

1° Presentation of the task

At this stage, the task to be carried out is presented to the participants, along with instructions describing the nature of the work to be carried out individually and/or in groups. In short, it's important at this stage to specify or clarify the instructions as far as possible. In simple terms, this stage consists of giving the learners the information they need, initiating the activity and clarifying the instructions for carrying out the task.

It's important to add that it's at this stage that the expositive and active methods can be combined. Indeed, when presenting a task, particularly when introducing a new notion that is particularly unfamiliar to learners, the teacher has the latitude to resort to a beautiful exposé. So, as modern didactics acknowledges, the marriage between the transmissive model and active methods is indicated in the teaching-learning process.

2° Individual and/or group work

In the light of the instructions given in the above step, students carry out the work they have been asked to do, individually and/or in groups. In the case of group work, it's almost inevitable that learners will complement each other's work through discussion and exchange. Very often, mistakes made or difficulties encountered by students during individual work are resolved during group work.

3° Restitution or pooling

After the students have worked in groups, where they have had the opportunity to compare and contrast their opinions, it's usually time to pool their findings. This involves presenting the results of each group's work in front of everyone. It should be added that, in addition to the results of the groups' work, the teacher is required to make partial fixations on each of the students' contributions. It's a very important moment for fruitful exchanges between learners, under the aegis of the teacher, and where students really learn.

As this is the privileged moment for the discovery of knowledge, Grêt insists that sufficient time must be devoted to this stage if didactic fertilization is to take place.

4° Summary

Facilitated by the teacher, the synthesis is the presentation of the solution to the problem posed at the outset. Very often, this is the moment when students become aware of the work they have done, i.e. they realize whether the solution they have found is true or false in relation to the problem.

What's more, these four steps are general to all the PAP methods we can use. However, each method (Group I performance, Group II performance, etc.) approaches them in its own particular way.

Whatever the participative, active, ... character advocated by the competency-based approach dear to Xavier Roegiers (APC), Gérard Mubangu (2015, p. 164) enables us to clearly differentiate it from Christian Grêt's approach. He proposes the following table:

Table 4. Comparison between PAP and APC

Active, participatory teaching	Competency-based approach
Elements of similarity	
✚ Constructivism and socio-constructivism ; ✚ Skills development: knowledge, know-how, interpersonal skills, etc.	
Elements of dissimilarity	
✚ Applicability to both general and vocational education ; ✚ Lessons are always given in a group,...	✚ Particularly applicable to vocational education; ✚ Lessons are sometimes given in groups or individually

1.4. OVERVIEW OF EVALUATION FORMS

In addition to the fact that evaluation is an inseparable process from the teaching-learning process, the pedagogical approach that drives the teaching-learning process ipso facto imposes itself on the form of evaluation. Referring to the *development of a variety of internal assessments*, the CNESCO (2014, p. 15) points out that some countries prefer *traditional forms of assessment* (as in France), such as written homework, the content of which is left to the teacher's discretion, while other countries advocate or require, in official texts, the use of *new forms assessment* asEngland and Quebec), such as self-assessment, peer assessment or individualized student monitoring. With this in mind, Annick Lavoie *et al* (2012, p. 6) point out that it is possible for there to be *group* or *individual assessment*, especially as some favor group assessment while others advocate individual assessment and see no conflict between individual assessment and cooperative pedagogy. Once again, in the absence of clearly established and proven principles in this area, it will be up to the teacher to choose between group and individual evaluation, depending on the cooperative and disciplinary skill development objectives he or she is pursuing and the activities he or she is implementing

While taking these parameters into account, in the present work, forms of evaluation are to be grouped into two camps: evaluation according to the traditional approach and according to the PAP approach.

1.4.1. Traditional pedagogical assessment

As this pedagogy emphasizes competitiveness, selectivity, etc., the form of assessment that applies is none other than individual assessment, which, according to Paul Marie Konsebo and Sekhna Sylla (2015, p. 25), consists of the teacher interviewing each student in the group/class by name or individually, in order to monitor his or her progress at the end of the activities carried out.

1.4.2. Valuation applicable in the PAP

As previously mentioned, PAP has its origins in the constructivist theories of Jean Piaget and Lév Séménovitch Vygotsky, and the assessment models applicable to this approach take this into account. This is why there is self-evaluation (cf. Piagetian constructivism) and cooperative evaluation (cf. Vygotskian socio-constructivism).

1.4.2.1. Self-evaluation

For Françoise Campenale (2001, p. 23), this is an internal evaluation by the subject of his or her own action and what it produces. To this we can add Nunziati's reflexion, quoted by Campenale (Ibid., p. 12), that this evaluation must be defined as a dialogue between oneself and oneself.

As Broadfoot quoted by Olivier Rey and Annie Feyfant (2014, p.28) points out, *self-assessment, therefore, is not just an assessment practice, it is also a learning activity. It is a way of encouraging students to reflect on what they have learned, to look for ways to improve their learning, and to plan what will enable them to progress as learners and achieve their goals. [...] As such, it includes skills in terms of time management, negotiation, communication with teachers and peers and self-discipline, in addition to reflexivity, critical thinking and evaluation.* This is identical to Linda Allal's idea, quoted by Campenale (op. cit., 12), for whom self-evaluation is not evaluation in the true sense, but a metacognitive reflection that triggers self-regulation on the part of the student. It enables the student to answer the following questions: what do I know, what can I do, how do I do it, what can I change?
To this end, Campenale schematizes self-evaluation as follows:

Figure 7. Self-evaluation

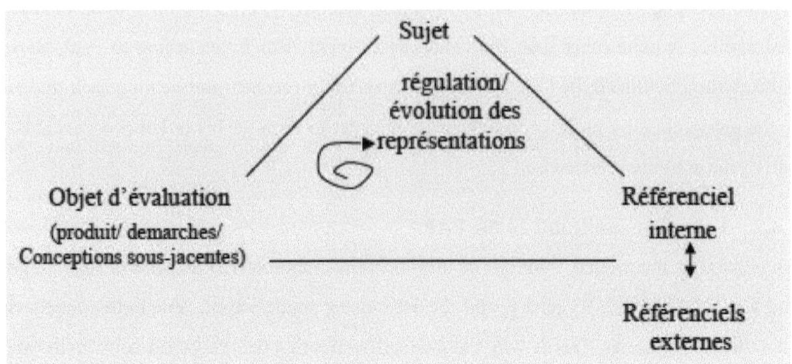

Let's remember that this is part of formative evaluation, since it enables individual learners to become aware of their own progress.

1.4.2.2. Cooperative or interdependent assessment

Paul Marie Konsebo and Sekhna Sylla (op.cit, p.25), believe that this was set up to reinforce the spirit of mutual aid and solidarity within the group, with the teacher organizing exercises for the group from time to time. He or she may also interview learners on behalf of the group. It is said to be interdependent, as El-Habib Berra (2014, p. 39), Slavin, Lejk *et al.* and Isabelle Plante (2012, p. 260) point out, not only because it is based on the quality of work done together, but also because the work of each member affects the feedback received by all group members.

This assessment requires students to be involved in correcting and marking their own homework (or in pre-correcting it themselves).

In this sense, Pepper, quoted by Olivier Rey *et al* (op.cit, p. 29), has this to say: *from a 21st century skills assessment perspective, cooperative work is, it should be remembered, an essential dimension: encouraging collective activities in such a crucial field is a way of getting students used to working together in a sector that seems spontaneously marked by individual competition.*

Davis, quoted by Dumas & Huguet and echoed by Olivier Rey *et al* (Ibid.), reinforces this view, arguing that students feel less successful in selective schools. These observations produced the concept of the school self and the contrast effect; "*it's better to be a big fish in a small pond*" than the other way round.

With this in mind, Gérard Mubangu (2015, p. 164), points out that group work can also be used to assess learning achievement. All the more so as it is one of the strategies implemented by the PAP.

In this way, the author proposes a comparison of forms of evaluation, adapting them to the pedagogical approach.

1.4.3. Comparison of individual versus group assessment

As mentioned above, the characteristics and elements of these two types are identified in the table below:

Table 5. Differentiation between group and individual evaluations

Team evaluation	Individual assessment
Interaction between students	Feeling of being top of the class
A spirit of cohesion	Personal interests
Socio-cognitive situations	Perception of self, not group
Interdependence game	Sense of competition
Spirit of solidarity	Feeling of surpassing others
Cooperative spirit	Group isolation spirit
A spirit of complementarity	Individualism
Trusting your partner	Etc.
Socio-constructivism	
Student collaboration	
Etc.	

Source: Gérard Mubangu (Ibid., p.166)

1.5. PREVIOUS STUDIES

Since nothing falls from heaven like *manna to the Israelites* in the desert, it's clear that the present study requires us to draw on several other works for reference:

1. François Le Menaheze (2002), *Coopérer ... pour apprendre ! La coopération entre élèves, la médiation de l'enseignant : un processus de co-construction des savoirs,* Unpublished Master's thesis in Educational Sciences, University of Nantes . Has the idea that students can no longer face knowledge and the teacher

alone, but are part of a process of co-construction with others (students and teacher) gained ground? With this question in mind, François predicts the answer above: cooperative work produces positive effects on the cognitive interactions of students in the classroom, and that these interactions, within the framework of the cooperative classroom, enable students to co-construct knowledge.

The aim of his work was to observe how students cooperate and interact in the classroom, and to better understand the mechanisms involved in the process of building knowledge through cooperation between students. To achieve this, he began with observation and an open-ended questionnaire. After various analyses, he came to the unquantified conclusion that group work is more effective than individual work.

2. Dunia quoted by Gratien Mokonzi (2015b, pp. 37-38) who, for his DES dissertation, studied the impact of teamwork, one of the PAP strategies, on student learning. Using a time-series or longitudinal experiment, he collected data for his dissertation at Collège Maele in Kisangani. To this end, he administered a 10-question mathematics test. The test concerned the chapter on sets in secondary 1ère, and he administered it 3 times before the chapter was taught, and again after it had been taught. His sample consisted of 20 students, 10 per class. For him, the 1ère A pupils were the experimental group, and those in 1ère B the control group.

The average results obtained by students in these two classes are as follows:

- Pre-test:
  Pre-test1: 1st A (0.6) and 1st B (0.7)
  Pre-test2: 1st A (1.6) and 1st B (1.6)
  Pre-test3: 1st A (2.6) and 1st B (2.5)
- To the post test:
  Post test1: 1st A (6.6) and 1st B (2.8)
  Post test2: 1st A (7.6) and 1st B (2.9)
  Post test3: 1st A (8.0) and 1st B (3.2)

  By observing these averages, and after statistical testing, he realized that the two classes were initially at the same level - curiously enough, just after post-test 1, a very significant difference began to emerge, in favor of the experimental group.

This led him to conclude that teamwork is more effective than individual work, and that this effectiveness is long-lasting, especially as the averages achieved by the 1st A group continue to improve.

3. Gérard Mubangu Wa Kapala (2014), Rendement des méthodes de la PAP et de la pédagogie traditionnelle dans les écoles primaires de Bagira au TENAFEP 2011-2012, in *Cahiers du CERUKI, Nouvelle série* n° 45, CERUKI, Bukavu.

   The aim of this article was to detect the results of PAP and traditional pedagogy methods on the performance of Bagira elementary school pupils in the 2011-2012 TENAFEP, and the results of interactions between pupils and between them and the teacher during didactic sequences.

   To this end, the following questions guided the research:

   - How do the 6th grade pupils in Bagira elementary school perform in the TENAFEP 2014-2012 session using PAP or MPT?
   - What are the results of interactions between students during a didactic sequence
   - What are the outcomes of teacher-student interactions during a didactic sequence?

   To answer these questions, he used a pre-experimental method with two groups. Observation and documentation were used to collect the data. As for analysis, he calculated averages, percentages and standard deviations, and a comparison of averages using Student's t-test brought this to a close.

   Thus, with a calculated t of 6.200, which is higher than the critical t of 1.87696, the H_0 was rejected, allowing us to conclude that the PAP methods significantly modify the results of TENAFEP students in the Bagira commune.

It's worth pointing out that not only these works, but others have also influenced the present in one way or another. In other words, this list is not exhaustive.

Chapter two: METHODOLOGICAL FRAMEWORKS

This chapter deals with the methodology used in this work to achieve its objectives. It covers the study population, the sample, the methodology used to collect the data and, finally, the way in which the data were analyzed.

2.1. STUDY POPULATION

As Gratien Mokonzi (2013, p. 38) puts it, the study population or parent population is a set of all subjects with specific characteristics related to the study or research objectives. The author adds that it can be finite or finite.

As school statistics for the current year are not yet available, this work finds itself in a situation where the study population is infinite. Clearly, this population is made up of secondary school pupils in the Protestant network in South Kivu province, since it is the ECPs who benefit from PAP training.

2.2. SAMPLE

Gilbert de Landsheere's (1972, p. 251) idea that *to sample is to select a limited number of individuals, objects or events whose observation enables us to draw conclusions (inferences) applicable to the reference set (universe) within which the choice was made.*

For this purpose, Institut Imani Panzi was chosen because, as Joseph Barafumwa (2015, p. 5) points out, this school is one of two secondary schools reputed to have regularly applied the PAP methods of all the city's schools that have benefited from the said training. And given that there are still some teachers in this school who are not trained in PAP, and that the two approaches need to be compared, this is the school that allows conclusions to be drawn that are applicable to all ECP/SK.

It's important to point out that the stratified random sampling technique was used to select a sample of 32 6th secondary school pupils, 16 from each class, including 16 from MP and 16 from CA at the school. The procedure was as follows: once we had the lists for these two classes, the pieces of paper were numbered according to the number of pupils in each class, folded and placed in two empty chalk boxes, one for the MP pupils and the other for the CA pupils. This operation made it possible to randomly draw up to 16 pieces per class. It was by these numbers that the names of the subjects concerned were consulted on the lists.

The graph below summarizes the source population.

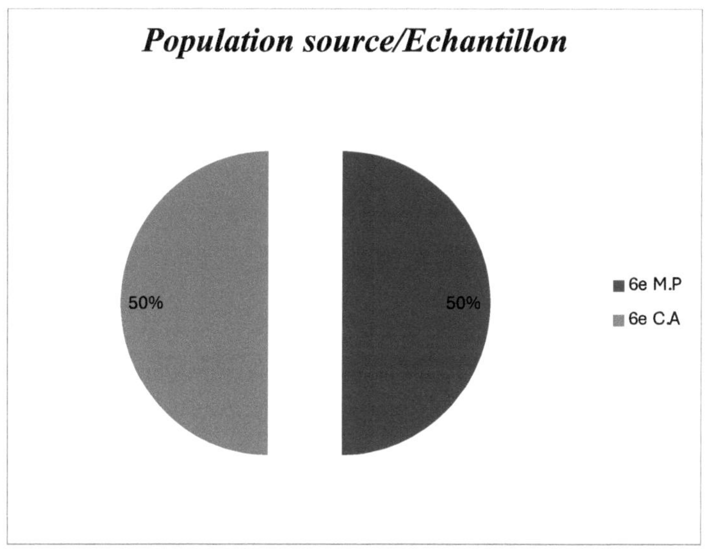

Figure 1. Sample

2.3. METHODOLOGY

It's worth pointing out that the documentary technique and the experimental method have guided the collection of data for this work at all levels.

In this way, we discovered a number of texts that Christian Grêt (2009, pp. 46-47) has always used to train PAP teachers. It is from this list that this work has drawn on one of these texts, in which a French dictation test was administered to 6th secondary school pupils at the Institut Imani Panzi. This test was administered in two phases, the first of which was considered the *pre-test* and the second the *post-test.* The pre-test was administered individually to both classes, at the same time and in the same room, under the supervision of the teachers and the researcher. The post-test was carried out in two different ways: *individually* for the control group and *in a group* for the experimental group.

This French dictation test was chosen because it is the one that Christian Grêt (Ibid., p. 44) has always used to demonstrate to teaching practitioners that *group work* is increasingly more effective than individual work. Since he administers it to teachers of all levels (nursery, primary and secondary), students in educational science and school heads, for this work a paragraph of the chosen text was sufficient to be administered to 6th secondary school pupils.

To this end, this work has proceeded by means of quasi-experimentation according to the *basic experimental design* in order to collect empirical data. That is to say, given that the experiment will be applied in a situation where there are two groups, the starting level will have to be tested for both, and after the treatment, which in the context of this study is the fact of *having the pupils work together in small groups*, these pupils will then take the same test a second time to see the progress made by each group/class. Schematically, the situation is as follows:

Figure 8. Basic experimental design

O_1	X	$O_2 \leftrightarrow$ Experimental Group
O_3		$O_4 \leftrightarrow$ Control group

Source: Gratien Mokonzi (2015b, p. 32)

Caption:

- O_1: Observation n°1: pre-test in relation to the experimental group ;
- O_2: Observation n°2: post-test compared with experimental group
- X: Processing which represents *group work* for the Experimental Group
- O_3: Observation n°3: pre-test versus control group
- O_4: Observation n°4: post-test versus control group

It is important to note that Observations 1 and 3 are the pre-test, while 2 and 4 are the post-test.

It's also worth pointing out that the French teacher in all these $6^{(th)\ classes}$ is not yet trained in PAP. In view of this, a request was made to the school management and the response was favorable, as they agreed that the teacher in charge of the same course in secondary 1ère, who was already trained in the said approach, could do some didactic sequences with the students in the Experimental Group and administer the post to the same students following the form of evaluation corresponding to the socio-constructivist model.

As far as marking was concerned, it was agreed that, since the text was not extremely long, one mistake equated to minus one (-1), given that these students had a "slightly high level". And the test was graded on 10 points. Immediately after the post-test, an overall correction (teachers, students and researcher) took place.

To be explicit, it is important to point out that "Group 1 performance" was applied for these purposes, and so the procedure was as follows:

After the teacher had clarified the instructions, he dictated the text and each student worked individually. He told these students to work in pairs as they sat on the bench, and then he told them to work in fours, after which the secretaries of these different groups went ahead to write down the results their groups had achieved on the TN. After the other group members had affirmed the loyalty of their secretaries, the teacher and students all identified the mistakes made by each group. In short, it was the results of the work done in groups of four that were the subject of the post-test of the present study.

2.4. DATA PROCESSING

For Humblet (1980, p. 117), this stage consists of summarizing, synthesizing and classifying the data collected, taking into account the various facets of the problem, notably the description of origins, causes and consequences. Léon Nguapitshi (2012, p. 40) adds that data analysis can be carried out manually or using a computerized tool.

SPSS software was used to classify, summarize and/or synthesize the data collected in the field.

2.5. DATA ANALYSIS

After the counting phase, the data must be analyzed

Taking into account the experiment carried out, it is essential to carry out a two-way analysis: *inter-group*, to determine the difference caused or not by the treatment, and *intra-group*, to identify the changes brought about by each group. The procedure is as follows:

- Inter-group comparison
    - ✓ O_1 and O_3 averages to determine the equivalence level at start-up;
    - ✓ O_2 and O_4 averages to determine treatment effects
- Intra-group comparison :

- ✓ Averages of O_2 and O_1 to determine the change brought about by treatment (X) in the experimental group;
- ✓ O_4 and O_3 averages to determine the degree of evolution of the control group.

Since we're comparing the experimental group with the control group, we need to apply Dunnett's *t-test*. The formula is as follows

$$t_d = \frac{|\overline{X}_j - \overline{X}_o|}{\sqrt{\frac{2\,ve}{n}}}$$

Caption:

- ❖ $\overline{X}_j$ Experimental group average ;
- ❖ $\overline{X}_o$ Control group average ;
- ❖ 2 : number of groups ;
- ❖ *Ve*: error variance or simply variance ;
- ❖ *n*: number of subjects.

Chapter Three: PRESENTATION, DATA ANALYSIS AND INTERPETATION OF RESULTS

As its title suggests, this chapter includes sections on data presentation, how the data was analyzed or processed, and the results obtained.

In a situation where the sample is greater than 30, statistical standards recommend that we group the data. To do this, we use the tables and graphs provided by SPSS software, and the data are presented as follows:

3.1. PRE-TEST RESULTS

3.1.1. *Experimental group*

Table 6. Frequencies for O_1

	Frequencies	Workforce	Percentage	Cumulative percentage
Valid	0.00	5	31.3	31.3
	1.00	4	25.0	56.3
	2.00	2	12.5	68.8
	4.00	3	18.8	87.5
	5.00	1	6.30	93.8
	6.00	1	6.30	100.0
	Total	16	100.0	

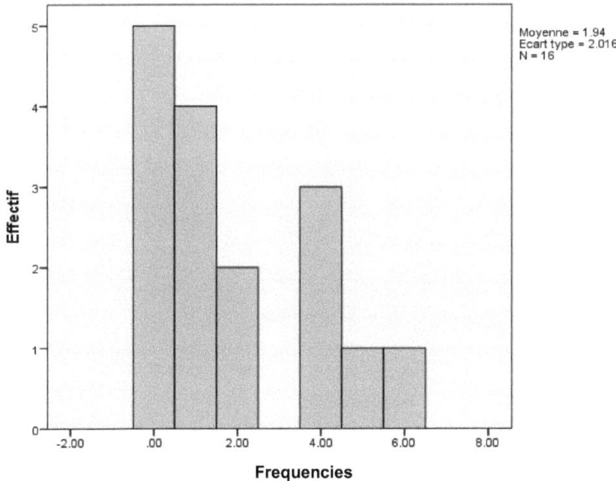

Figure 2. Pre-test results for experimental group

The table and graph show that of these students, 5 (31.3%) scored 0/10, 4 (25%) scored 1/10, 2 (12.5%) scored 2/10, 3 (18.8%) scored 4/10, 1 (6.3%) scored 5/10 and finally 1 scored 6/10. These results give an average pass rate of 1.94 and a standard deviation of 2.016.

3.1.2. *Control group*

Table 7. Frequencies for O_3

Frequencies		Workforce	Percentage	Cumulative percentage
Valid	0.00	6	37.5	37.5
	1.00	1	6.30	43.8
	2.00	3	18.8	62.5
	3.00	3	18.8	81.3
	4.00	1	6.30	87.5
	5.00	2	12.5	100.0
	Total	16	100.0	

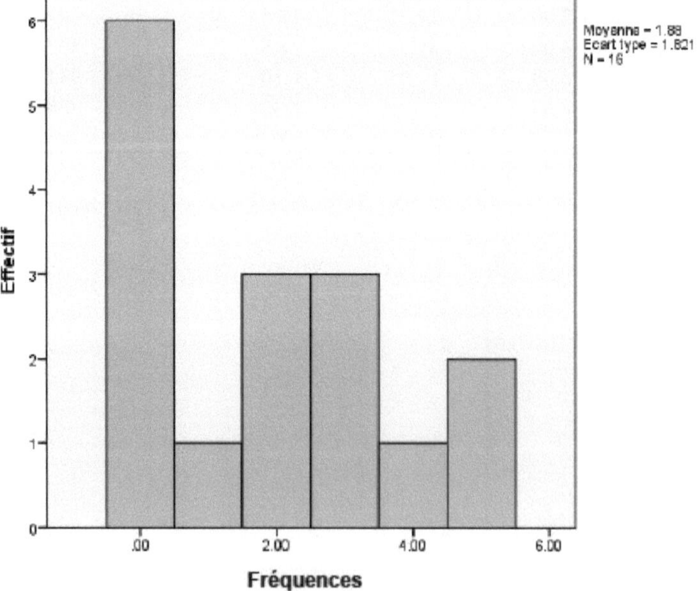

Figure 3. Pre-test results for the control group

The table and graph show that 6 students (37.5%) scored 0/10, 1 student (6.3%) scored 1/10, 3 students (18.8%) scored 2/10, 3 students (18.8%) scored 3/10, 1 student (6.3%)

scored 4/10 and 2 students (12.5%) scored 5/10. The average pass was 1.88 and the standard deviation 1.821.

In summary, the pre-test situation is shown in the table below:

Table 8. Descriptive statistics for the pre-test

Groups	N	Max	∑. Points	$\overline{X}$	Standard deviation	Variance	CV	Rdt.
Experimental	16	10	31	1.94	2.016	4.063	1.039	19.4
Control	16	10	30	1.88	1.821	3.317	0.969	18.8
Total	32							

In relation to the above table, the sum achieved is 31 for the experimental group versus 30 for the control group, in addition to which, the average pass achieved by the experimental group is 1.94 versus 1.88 for the control group. In terms of indices of dispersion, the standard deviation was 2.016 for the experimental group versus 1.88 for the control group, the variance was 4.063 for the experimental group versus 3.317 for the control group, and the CV was 1.039 for the experimental group versus 0.969. In view of these CVs, it is wise to say, according to the interpretation scheme proposed by Gratien Mokonzi (2015c, p. 47), that there is no homogeneity in the two classes. All the more so as in both groups the CV is well over 0.30. In short, since the CV in all these groups is above 0.30, the disparities between the students considered are very pronounced. Finally, the performance of the experimental group was 19.4%, compared with 18.8% for the control group. When interpreting performance, it is important to refer to the 50% standard which, in our country, is the starting point for success. And since both classes failed to achieve this percentage, they must both be judged as *ineffective* in relation to French dictation at pre-test

3.2. POST-TEST RESULTS

3.2.1. Experimental group

Table 9. Frequencies for O_2

Frequencies		**Workforce**	**Percentage**	**Cumulative percentage**
Valid	5.00	4	25.0	25.0

6.00	12	75.0	100.0
Total	16	100.0	

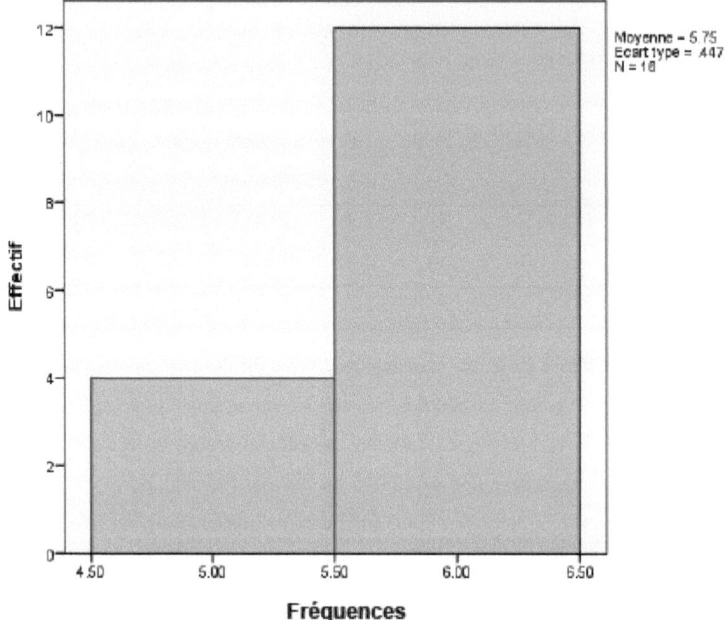

Figure 4. Post-test results for experimental group

From this table and graph, it should be noted that 4 students (25%) scored 5/10, compared with 12 (75%) who scored 6/10. The average score for this group was 5.75, with a standard deviation of 0.447.

3.2.2. Control group

Table 10. Frequencies relative to O_4

Frequencies		Workforce	Percentage	Cumulative percentage
Valid	0.00	7	43.8	46.7
	2.00	3	18.8	66.7
	4.00	3	18.8	86.7
	5.00	1	6.30	93.3
	8.00	1	6.30	100.0

	Total	15	93.8
Missing	Missing system	1	6.30
	Total	16	100.0

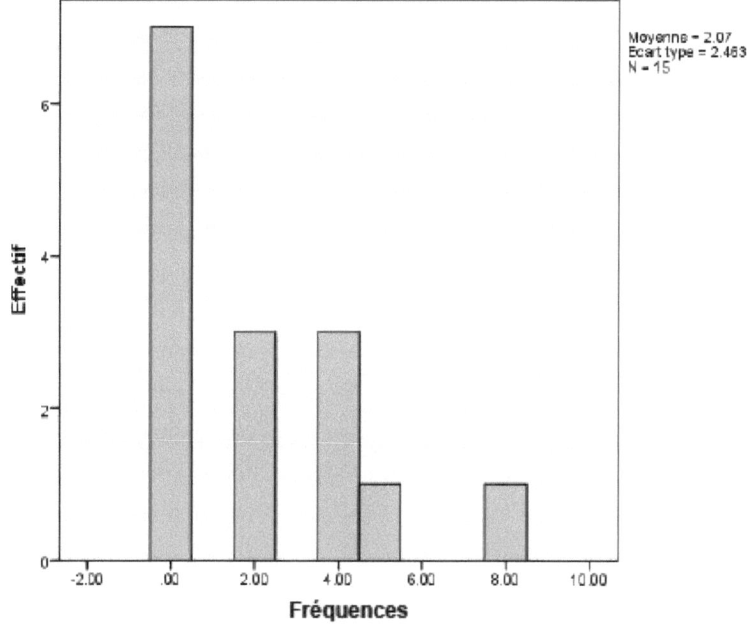

Figure 5. Post-test results for the control group

As far as this observation is concerned, 7 students (43.8%) scored 0/10, 3 (18.8%) scored 2/10, the others 3 (18.8%) scored 4/10, 1 (6.3%) scored 5/10 and finally 1 scored 8/10. The average pass was 2.07 and the standard deviation was 2.463.

To sum up, the reality resulting from the post-test is shown in the table below:

Table 11. Descriptive statistics for the post-test

Groups	N	Max	∑. Points	$\overline{X}$	Standard deviation	Variance	CV	Rdt.
Experimental	16	10	92	5.75	0.447	0.200	0.078	57.5
Control	15	10	31	2.07	2.463	6.067	1.189	20.7
Total	31							

For this second stage, there is a slight discrepancy with the sample, especially as one of the students in the control group didn't turn up. For this reason, instead of 16, only 15 subjects from this group are considered. As the above table shows, the important thing to remember is that the experimental group achieved a sum of 92 compared with 31 for the control group, giving the following averages: 5.75 for the experimental group and 2.07 for the control group. As for the dispersion indices of interest to us, let's say that the standard deviation for the experimental group is 0.447 versus 2.463 for the control group; the variance is 0.200 for the experimental group and 6.067 for the control group; and finally, the CV for the experimental group is 0.078 and that of the control group is 1.189. As the CV of the experimental group is below 0.15, there are no disparities between the students considered, but in the case of the control group, small disparities are observed. This means that we can say that the trend has just been reversed in favor of the experimental group, which is becoming increasingly homogeneous compared with the control group. As for yield, the experimental group achieved a post-test yield of 57.5%, compared with 20.7% for the control group. This means that it is the experimental group that is worthy of being judged *effective*, compared with the control group, which continues to keep its yield well below 50%. The control group is therefore considered *ineffective* in this situation.

3.3. PRESENTATION AND INTERPRETATION OF RESULTS

Now that all the data has been presented, classified and summarized, it's time to move on to analysis and processing.

It should be remembered that Dunnet's t-test will be used to carry out the analyses for the various comparisons recommended to us by the experimental design that led to this research. These comparisons are as follows:

3.3.1. Intergroup comparisons

These comparisons involve identifying the differences between the experimental and control groups, depending on when the test was taken. Hence a comparison of means for the pre-test and another for the post-test.

3.3.1.1. Comparison of pretesting effects (0_1 - 0_3)

From the calculations we made, the t_d we found is 0.088. As this *calculated* $t_{(d)}$ is lower than the *critical* $t_{(d)}$ at the 5% threshold (2.04), we can see that the $t_{d\ (cal)}$ falls within the acceptance zone. In simple terms, as the $t_{d\,(cal)}$: 0.088< $t_{(d\,(tab)).05}$, we accept the H_0. Accepting the null hypothesis means that there is no significant difference between the experimental group and the control group in the pre-test. Or quite simply, the starting level of 6th CA and 6th MP students is homogeneous.

3.3.1.2. Comparison of treatment effects (0_2 - 0_4)

After our calculations, we found a t_d of 5.795. Since this *calculated* $t_{(d)}$ is greater than the *critical* $t_{(d)}$ at the 5% threshold, we reject the null hypothesis. In simpler terms, as $t_{d\,(cal)} > t_{d\,(tab).05}$, we reject the H_0.

Rejecting the H_0 here means that there is a significant difference between the experimental group and the control group. Thus to say that the treatment (group work) has a positive impact on the result of the experimental group.

3.3.2. Comparisons of intra-group effects

As far as this point is concerned, the aim is to analyze the evolution of each group, i.e. to establish a relationship between the results of the pre-test and those of the post test in each group independently of the other.

3.3.2.1. Comparison of experimental group effects (0_2 - 0_1)

Given that the calculated $t_{(d)}$ of 7.33 is higher than the critical $t_{(d)}$ at the 5% threshold, it makes sense to reject the null hypothesis and accept the alternative hypothesis, especially as the value 7.33 falls within the rejection zone of the curve.

This third comparison shows that H_0 is still rejected, meaning that the level of the experimental group in the post-test has improved significantly compared to the pre-test.

3.3.2.2. Comparison of control group effects (0_4 - 0_3)

Our calculations show that t_d is 0.244. This result is lower than the critical $t_{(d)}$ at the 5% threshold, so we must accept the null hypothesis, especially as the calculated value lies within the acceptance zone of the curve.

Accepted here, $H_{(0)}$ means that there are no pronounced differences between the average pass rate achieved by this group in the pre-test and that of the test station.

Chapter four: DISCUSSION OF RESULTS

Through comparisons of averages, the aim was to detect which of these two modes of work had a positive impact on learner success compared to the other. In other words, the aim of these analyses was to ascertain the influence of the form of assessment on student performance, by contrasting group assessment, a strategy used in the PAP approach, with individual or selective assessment, which is often used in the PT approach.

As a result, this chapter is structured around the following points:

4.1.EXPERIMENTAL GROUP

As indicated in section (3.3.2.1.), the calculated t_d is well above the critical $t_{(d)}$ at the 5% threshold. Such a situation leads us to conclude that there has been an evolution by the students in this group.

Figure 9. Group work

In the image above, students in the 6th MP class at Institut Imani Panzi are working in a climate of mutual aid, cooperation,...

Drawing on various laboratory studies, Jean Paul Roux (n.d), points out that *interactions between peers* in problem-solving situations play a *constructive* role in individual cognitive skills. He adds that, whatever the level of the collaborator (lower, equal or higher), the dynamic of exchanges between students draws significant cognitive benefits.

To this end, Céline Buchs *et al* (2004, p.172) state that: *social interdependence represents a situation in which individuals present a common goal, and the outcome of each is affected by the outcome of others.*

Whatever the apprehension of students and difficulties inherent in evaluating group work, Mello, as quoted by Anastassis Kozanitis (2005b, p. 1), presents five reasons in favor of the use of group work and its evaluation. The author believes that this type of work enables students to :

- Acquire an accurate understanding of group dynamics;
- Work on a larger scale than individual work;
- Developing interpersonal skills ;
- Being confronted with different points of view and
- Better prepare students for the demands of the real world.

Let's end this point with Perret-Clermont, quoted by François Le Menaheze (op. cit., p. 93), for whom any pedagogical practice tending to individualize teaching should be jointly based on an *intensification of social interactions between students*

4.2.CONTROL GROUP

By the result presented in point (3.3.2.2.), the calculated $t_{(d)}$ being located in the table's acceptance zone, we are obliged to accept the null hypothesis. This means that there is no significant difference between the pre-test and post-test averages for these students. In other words, there has been no evolution in the performance of the students in this group.

Figure 10. Individual work

For Merle, quoted by Olivier Rey *et al* (op.cit., pp. 10-11), *school is a sort of huge sorting and labelling station for people according to their skills.* The merit society is therefore inevitably a society of measurement. Maulini, quoted by these authors, points out that this way of looking at evaluation stems from the Jesuit pedagogy, which, for the author, *is characterized by discipline, repetition* and *perpetual competition between pupils.* In this way, the authors draw a parallel with the elitist model deployed under the influence of the Chinese mandarinate, marked by competitive meritocracy for the selection of the Empire's administrators. Ranking becomes an essential criterion for judging a student within a given hierarchy (class, group, school,....).

4.3.COMPARISON BETWEEN EXPERIMENTAL AND CONTROL GROUPS

Since the treatment is to be judged primarily by the post-test, it's clear that the post-test is an important part of the decision-making process. And so it is that point (3.3.1.2.) presents the result that calculated $t_{(d)}$ is greater than critical t_d. In terms of difference, this result proves that the students who worked in groups achieved a higher performance than their colleagues who worked traditionally or individually.

Having also compared the way pupils work, Baudrit, quoting François Le Menaheze (op.cit, p. 104), writes as follows: *Many children fail when they work alone; they do better in the presence of a peer.* He adds that the dyadic situation is therefore likely to offer a heterogeneous interactive space, a space conducive to the encounter of different forms of thought.

Luxembourg, quoted by Serres and echoed by Olivier Rey *et al* (op.cit., p. 8), is bitterly *convinced that most teachers believe that pupils learn only under the threat of bad marks, deferments and repeats.* However, Barrère, quoted by these authors, notes that many teachers experience assessment as a *chore*, symbolized by copy packages, and as an area of activity that needs to be limited so that it doesn't take precedence over other teaching activities.

Thus, in the view of Rémond (2008), quoted by Olivier Rey *et al* (op.cit, p. 7), *assessment is only valid if it contains information that makes further learning possible, and if teachers use this information to adjust their teaching.* Beyond the learning and adjustment aspect mentioned by this author, Smith, quoted by Natalie Younes (2015, p. 7), adds that the important subsequent aspect that can be fostered by a good assessment is the professional aspect.

For this reason, it's obvious to tell teachers who are thinking along these lines that today, through the evaluation process, it's time to see things differently, in light of the new philosophy being pursued not only by the Congolese education system, but throughout the world, according to which we need to train men who are useful not only to themselves, but also to society, something which allows us to affirm that man does not exist alone. In other words, whatever his qualities, assets, etc., he is infinitely in need of someone else (big or small). In short, it's time to enable children to be cooperative, because tomorrow they'll be in the professional world where they'll be called upon to work with others.

To conclude, Maria-Alice Médioni, quoted by Frédéric Arthur (n.d), Claude Ernest Njoya and Gratien Mokonzi (2016), states that the aim of group work is not to answer a simple question or one that can be solved individually. On the contrary, the aim of this type of work is to open up avenues of solution, to put forward hypotheses, something that can only be done with others if we want to have several avenues and hypotheses that are as varied as possible. To put it plainly, group work should only be applied to complex notions, and not to small or simple ones that students can work out on their own.

CONCLUSION

The present study, which compared group assessment using the "socio-constructivist model" with individual or selective assessment of 6th year secondary students at the Institut Imani Panzi in the French dictation test, has come to an end. At the outset, the study was limited to the following question: *does the use of the socio-constructivist model (PAP or group work) in the assessment process have a significant impact on learners' academic performance?* To this end, the aim was to understand how effective PAP practice is in the process of assessing students' academic achievements. Or, quite simply, to understand the influence that MPAPs can have on learner performance. More specifically, the aim was to identify or understand, using the data collected, the necessity of the PAP that other education specialists attach to this new pedagogical approach. In short, the aim of this work is to publicize the importance of the PAP in today's teaching practices, not only in Bukavu but throughout the DRC if possible.

For this purpose, French dictation test was chosen as the working instrument for collecting practical data (field data), and this was administered to a sample of 32 pupils from two 6th secondary classes at the Institut Imani Panzi on two occasions. In short, documentary analysis by means of a quasi-experimental analysis provided this work with field data.

Once the data had been collected, comparative analyses of the means were carried out using Dunnett's t-test. The results of these analyses were as follows: statistically speaking, the means of success were 1.94 for the experimental group versus 1.88 for the control group, and these, using Dunnett's t-test, led to the result that t_d calculated (0.088) is lower than the critical t_d at the 5% threshold (2.04), and as this $t_{d\ (cal)}$ falls within the acceptance zone, it was normal to accept the null hypothesis. This meant that there was no significant pre-test difference between the two groups. Surprisingly, there was a significant difference in the post-test, with the result that the average pass rate for the experimental group was 5.75, compared with 2.07 for the control group. And whose calculated $t_{(d)}$ (5.786) is significantly higher than the critical $t_{(d)}$ at the 0.05 threshold (2.04).

As for intra-group comparisons, it has also been shown that the evolution is more pronounced in the experimental group than in the control group. In short, the class considered as the experimental group significantly improved its performance over that of the control group.

With these results proving an absolute advantage for the experimental group, the assumption becomes a reality, and the hypothesis that the use of PAP methods in the assessment process has a significant influence on learners' results is confirmed. In other words, applying the socio-constructivist model to students' work enables them to do better than when working alone.

In the light of the above, the following suggestions emerged:

- ★ We call on the Congolese government to put into practice what it has always said about its involvement in the application of PAP methods in Congolese schools, which have reached an impasse and for which these methods are an effective remedy.
- ★ The authorities of USK in general, and FPSE in particular, should also follow the lead of ULPGL, which, aware of the need for this approach, has expressed its demand for such training for its education science students, to ensure their initial training not only as advisors, administrators and/or teacher inspectors, but also as duly trained teachers.
- ★ CP-ECP/SK and its partners involved in the introduction of PAP in the DRC, not only to focus their attention on schools in towns and villages closer to the cities, but also to extend this approach to schools further away from the cities.
- ★ It's up to PAP-trained teachers and those who haven't yet been trained to strengthen their cohesion and continue to work in synergy, helping each other to bring the Congolese school to excellence, of which this approach is an important part.

In this section, it's important to paraphrase Gérard Barnier (op.cit, p. 3), even though our research tends to emphasize the PAP, which stems from the work of Piaget and Vygotsky, that there is no absolute way that is fundamentally better than another: it all depends on the objectives to be achieved, the content being worked on, the people with whom we work, the institutional conditions in which we find ourselves as teachers, and so on.
It is also imperative to say that the present study is not a gospel where one must say amen. This is to say that it has had certain weaknesses, some of which are: a small number of samples, a short experimental period, etc. These and other unmentioned elements may ipso facto influence the conclusions drawn from the present study. It is therefore essential

that these aspects be revisited in our future research, or that other researchers enrich our research with future studies.

Finally, it's worth pointing out that no one is unaware that there are imperfections at any level, and these can be observed in any human work. For this reason, the author invites his readers to be tolerant and, if possible, to send him constructive observations to enable him to improve. Thank you for your attention!

BIBLIOGRAPHICAL REFERENCES

Aboubaker N. A. (2009), *L'évaluation scolaire : d'une conception à l'autre,* [online] at http://edufle.net (page consulted on November 20, 2015).

Arthur Fr. (n.d), *Le travail de groupe*, notes de synthèse non publiées, Académie de Nantes.

Barnier G. (n.d.), *Théories de l'apprentissage et pratiques d'enseignement*, unpublished conference, IUFM d'Aix-Marseille.

Barafumwa B. J. (2015), *Rapport succinct du suivi d'applicabilité de la PAP dans les écoles de la Ville de Bukavu*, Unpublished report by CP-ECP/SK.

Berra El-Habib (2014), *Travail de groupe dans l'apprentissage du FLE au secondaire : techniques et enjeux*, Unpublished Master's thesis, University of El-Oued.

Bercier-Larivière M. and Forgette-Giroux R. (1999), L'évaluation des apprentissages scolaires : une question de justeesse, in *Revue canadienne de l'éducation* (pp. 169-182) n° 24, 2.

Bloche *et al* (1999), *Grand dictionnaire de la psychologie*, Paris, Larousse.

Campanale F. (2001), *Quelques éléments fondamentaux sur l'évaluation*, unpublished course, IUFM Grenoble.

CNESCO (2014), L'*évaluation des élèves par les enseignants dans la classe et les établissements : réglementation et pratiques. An international comparison in OECD countries*, Paris: CNESCO.

Buchs C., Filiseti L., Butera F. and Quiamzade A. (2004), Comment l'enseignant peut-il organiser le travail de groupe ? in Gentaz E. & Ph. Dessus (Eds.), *Comprendre les apprentissages. Sciences cognitives et éducation* (pp. 168-183). Paris : Dunod.

Buchs C., Lehraus K. and Butera F. (2006), Quelles interactions sociales au service de l'apprentissage en petits groupes, in E. Gentaz & Ph. Dessus (Eds.), *Apprentissage en enseignement. Sciences cognitives et éducation* (pp. 183-199), Paris : Dunod.

Cusset P.Y. (2014), *Les pratiques pédagogiques efficaces : Conclusion des recherches*, Working paper n°2014-01, France Stratégie.

Delors J. *et al* (1998) *Education: un trésor est caché dedans,* Paris: Unesco

Dessus Ph. (2007), *Organiser le travail en groupe : Débattre et apprendre,* Séminaire d'analyse des pratiques d'enseignement/apprentissage non publié, IUFM Grenoble, [online] at http://www.upmf-grenoble.fr (page consulted on February 30, 2014).

De Landsheere G. (1972), *Introduction à la recherche en éducation*, Paris: Armand Colin Bourrelier.

El-Hage F. (2013), *Manuel de pédagogie universitaire*, Mission de pédagogie universitaire, Université Saint-Joseph de Beyrouth

European youth forum (2013), *Quality education policy paper*, Extraordinary General Assembly held in Thessaloniki, Greece from November 21-24, 2013.

Germain-Rutherford A. (n.d.), *Les modalités d'évaluation des acquis*, Unpublished notes, University of Ottawa, Canada Middlebury College, USA

Girault Isabelle (2007) *Théories d'apprentissage et Théories didactiques*, Cours de Master de didactique des sciences non publié.

Grêt C. (2006a), *Perception de la PAP en milieu africain par les enseignants et les autorités scolaires. Contribution à l'évaluation d'une expérience pédagogique*, unpublished DES dissertation, FPSE, UNIKIS.

Grêt C. (2006b), *Formation des enseignants à la PAP en milieu africain, Elaboration du portfolio pour la formation des formateurs*, Unpublished doctoral thesis, FPSE, UNIKIS.

Grêt C. (2009), *Le système éducatif africain en crise*, Paris: Harmattan.

Grêt C. (n.d), *Pédagogie active et participative*, [online] at http://www.pedagogie-active-partivipative.org (page consulted on 31/01/2014)

Hanna D., David I. and Francisco B. (Eds., 2010), *How do we learn? La recherche au service de la pratique*, CRIE - OCDE.

Houssaye, J. (1999), *Théories et pratiques de l'éducation scolaire (I), Triangle pédagogique*, 2(ème) édition, Berne: Editions scientifiques européennes.

Humblet J. E (1980), *Comment se documenter*, Brussels: Ed. Labour.

Inspection générale (2010), *Les grands courants de la pédagogie moderne*, Training module for secondary school heads and teaching advisors.

IREM de Toulouse (n.d), *Comment apprend-on ? Analyse de situations didactiques en mathématiques au collège*, unpublished notes.

Kambale N. D. (2016), PAP: Quelles exigences pour quel impact? in *l'éducation de l'E.E.C et de la C.B.C.A face aux défis et aux exigences des grands courants de la pédagogie dans un contexte d'insécurité et de dépravation des mœurs*, unpublished exchange workshop.

Konsebo P.M. & Sylla S. (2015), *Pédagogie des grands groupes,* Module d'autoformation pour les formateurs de formateurs du Burkina Faso.

Kozanitis A. (2005a), *Les principaux courants théoriques de l'enseignement et de l'apprentissage : un point de vue historique*, Bureau d'appui pédagogique de l'école polytechnique de Montréal.

Kozanitis A. (2005b), L'*évaluation du travail en équipe*, Bureau d'appui pédagogique de l'école polytechnique de Montréal.

Kozanitis A. (2015), *Pédagogie collégiale*, Vol 28, n°4.

Labédie G. & Amossé G. (2013), *Constructivisme ou socio-constructivisme?* [online] at http://gamosse.free.fr/socioconstruct/Rp70110.htm (page consulted on 19/01/2014).

Lavoie A., Drouin M. and Héroux S. (2012), La pédagogie coopérative : une approche à redécouvrir, in *Pédagogie collégiale* (pp. 4-8), Vol. 25, N° 3.

Le Coultre R. (n.d), *Socio-constructivism in the socio-constructivist model, how is knowledge constructed?* [online] at http://edutechwiki.unige.ch (accessed on 20/02/2016)

Le Menaheze F. (2002), *Coopérer ... pour apprendre ! La coopération entre élèves, la médiation de l'enseignant : un processus de co-construction des savoirs,* Unpublished Master's thesis in Education Sciences, Université de Nantes.

MINEPSP (2012), *Ecole amie des enfants,* Module de formation des enseignants sur école amie des enfants et la pédagogie active, Kinshasa: Direction des Programmes Scolaires et Matériel Didactique.

Ministère de l'éducation, du loisir et du sport de Québec (2006), *L'évaluation des apprentissages au secondaire. Cadre de référence*, © Gouvernement du Québec.

Mokonzi Gr. B. (2006), La pédagogie active et participative au chevet de l'école congolaise, in *Ecole démocratique,* Mise à jour du 20 avril 2006.

Mokonzi Gr. B. (2009), *De l'école de la médiocrité à l'école de l'excellence au Congo Kinshasa*, Paris: Harmattan.

Mokonzi Gr. B. (2013), *Méthodologie de recherche en sciences sociales et pédagogiques*, unpublished course, FPSE, USK/Bukavu.

Mokonzi Gr. B. (2015a), *Docimologie*, unpublished course, FPSE, USK/Bukavu.

Mokonzi Gr. B. (2015b), *Pédagogie expérimentale*, unpublished course, FPSE, USK/Bukavu.

Mokonzi Gr. B. (2015c), *Questions approfondies de la planification de l'enseignement*, Unpublished course, FPSE, USK/Bukavu.

Morandi Fr. & La Borderie R. (2006), *Dictionnaire de la pédagogie*, Paris: Nathan.

Mubangu K. G. (2014), Rendement des méthodes de la PAP et de la pédagogie traditionnelle dans les écoles primaires de Bagira au TENAFEP 2011-2012, in *Cahiers du CERUKI,* Nouvelles séries, n° 45, pp.156-161, Bukavu : CERUKI.

Mubangu K. G. (2015), Evaluation en groupe " Modèle socio- constructiviste " et évaluation individuelle des élèves de 4ème années de l'EPA Bwindi à l'épreuve de mathématique numération 2013-2014, in *Cahiers du CERUKI,* Nouvelles séries, n° 47, pp.159-167, Bukavu: CERUKI.

Mubangu K. G. and Birindwa M. H. (2015), Attitudes des enseignants à l'application des méthodes de la PAP dans les écoles primaires de Bukavu, in *Cahiers du CERUKI,* Nouvelles séries, n° 49, pp.119-132, Bukavu: CERUKI.

Nguapitshi K. Léon (2012), *Initiation à la recherche scientifique*, unpublished course, USK/Bukavu.

Njoya C. E. & Mokonzi Gr. B. (2016) *Mission d'évaluation du projet promotion du management scolaire et formation en PAP,* Unpublished preliminary report, CP-ECP/SK.

Plante I. (2012), L'apprentissage coopératif : des effets positifs sur les élèves aux difficultés liées à son implantation en classe, in *Canadian Journal of Education*, (pp. 252 - 283), 35, 3.

Rey O. and Feyfant A. (2014). *Évaluer pour (mieux) former. Dossier de veille de l'IFÉ,* n°94, September. Lyon: ENS de Lyon.

Roux J.P (n.d), *Socio-constructivisme et apprentissages scolaires,* [online] at http://dcalin.fr (accessed 20/12/2015)

Romainville M. (2002), *L'évaluation des acquis des étudiants dans l'enseignement universitaire*, Paris: Haut Conseil de l'Evaluation de l'Ecole.

Unesco (2014), *Teaching and Learning: Achieving Quality for All,* Paris: Unesco.

Unicef (2009), *Child-Friendly School Manual,* New York: Unicef

Younes N. (2009), Multi dimensionnalité de l'évaluation de l'enseignement universitaire dans la littérature anglophone, in Véronique Bedin, *L'évaluation à l'université. Evaluer ou conseiller?* Presses Universitaires de Rennes.

ANNEXES

Appendix 1.

Text chosen for dictation

Text by Rousseau, taken from Le *discours sur l'origine et le fondement de l'inégalité parmi les hommes* (1755) and reproduced by Christian Grêt (2009, p. 46).

First of all, it seems that men in this state, having no kind of moral relationship with each other, nor any known duty, could be neither good nor bad, and had neither vices nor virtues.

Let us conclude that wandering in the forests, without industry, without speech, without home, without war, and without liaison, without any need for his fellow men or any desire to harm them, perhaps even without ever recognizing any of them individually, the savage man, subject to few passions, and self-sufficient, had only the feelings and lights proper to this state; that he felt only his true needs, looked only at what he thought it was in his interest to see, and that his intelligence made no more progress than his vanity. If by chance he made a discovery, he could communicate it even less, as he didn't even recognize his children. Art perished with the inventor. There was no education, no progress; generations multiplied uselessly, each always starting from the same point, centuries passed in all the crudeness of the first ages, the species was already old, and man always remained a child

Appendix 2

Research certificate

UNIVERSITE SIMON KIMBANGU DE BUKAVU

N° 35, Avenue KIBOMBO/MAJOR- VANGU
Arrêté MINISTERIEL N°MINESURS/CABMIN/042/2008
E-Mail :univsimonk2005@yahoo.fr
www.uskbukavu.com

FACULTE DE PSYCHOLOGIE ET SCIENCES DE L'EDUCATION
ATTESTATION DE RECHERCHE N° [illegible]/USK/BKV/V - R/2016

L'(les) étudiant (e) (s) :

1. KYABENE MAKONGA
2.
3.
4.
5.

Régulièrement inscrit (e) (s) en DEUXIEME Année de Graduat/Licence au Département de SCIENCES DE L'EDUCATION Option : ADMINISTRATION ET INSPECTION SCOLAIRE à l'Université SIMON KIMBANGU de BUKAVU, USK/BKV en sigle, est (sont) autorisé (e) (s) de se rendre dans les institutions diverses (étatiques, para - étatiques et privées) suivantes :

1. INSTITUT IMANI (PANZI)
2.
3.
4.
5.

Pour des fins de recherche scientifique tel que le prévoit le programme d'enseignement dans le cadre de :

1. MEMOIRE DE LICENCE
2.
3.
4.
5.

Nous vous le (s) (la) recommandons à ce titre et vous prions de bien vouloir le (s) (la) recevoir et lui (leur) fournir les renseignements dont il (elle) (s) a (ont) besoin et lui (leur) venir en aide en cas de nécessité.

Fait à Bukavu, le 09/05/2016

Pour la Faculté,

Le Vice - Doyen.

Printed by Books on Demand GmbH, Norderstedt / Germany